TITLE VI

FISHERS JUNIOR HIGH SCHOOL
13257 CUMBERLAND ROAD
FISHERS, INDIANA 46033

A
19th CENTURY
FRONTIER FORT

Series Editor	David Salariya
Book Editor	Jenny Millington
Consultant	Dr John Huitson

Author:

Scott Steedman grew up in Australia and Prince George, a town in Western Canada that began as a fur-trading fort. He studied natural history at the University of British Columbia, Vancouver, and has edited many books on science and history for children.

Illustrator:

Mark Bergin was born in Hastings in 1961. He studied at Eastbourne College of Art and has specialised in historical reconstruction since leaving art school in 1983. He lives in East Sussex with his wife and daughter.

Consultant:

Dr John Huitson was born in North Shields in 1930. He studied at the University of Durham and then became Principal of Darlington College of Education. He has been Deputy Director and Director of Education at the American Museum in Britain, in Bath, since 1979.

Created, designed and produced by
The Salariya Book Co Ltd, Brighton, UK

First published in 1994
by Simon & Schuster Young Books

Simon & Schuster Young Books
Campus 400
Maylands Avenue
Hemel Hempstead
Herts
HP2 7EZ

ISBN 0-7500-1413-X

A catalogue record for this book is available from the British Library.

Printed and bound in Hong Kong by Wing King Tong Co., Ltd.

A
19th CENTURY
FRONTIER FORT

SCOTT STEEDMAN   MARK BERGIN

SIMON & SCHUSTER
YOUNG BOOKS

CONTENTS

INTRODUCTION	5
THE NEW WORLD	6
EXPLORERS AND SETTLERS	8
THE FIRST FORTS	10
THE WAY WEST	12
THE OREGON TRAIL	14
THE SITE FOR A FORT	16
BUILDING THE FORT	18
INSIDE THE FORT	20
PEOPLE OF THE FORT	22
FEEDING THE TROOPS	24
A SOLDIER'S DAY	26
TRADERS, FURS AND GOLD	28
THE PLAINS INDIANS	30
TRAPPERS AND TRADERS	32
INSIDE A PIONEER'S CABIN	34
A WOMAN'S DAY	36
BUILDING THE RAILWAY	38
THE INDIAN WARS	40
A FRONTIER TOWN	42
FORT FACTS	44
FAMOUS FRONTIER FOLK	45
GLOSSARY	46
INDEX	48

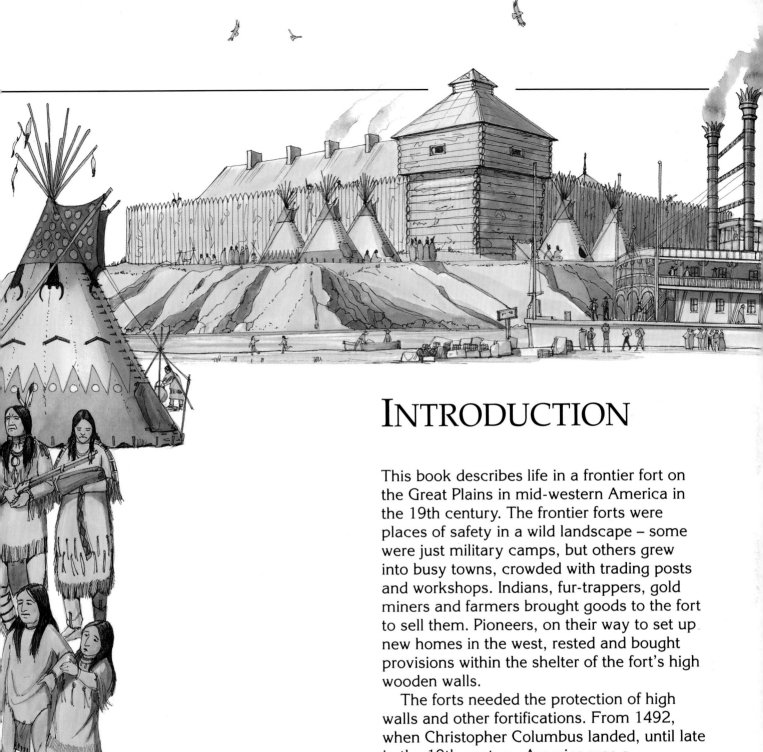

INTRODUCTION

This book describes life in a frontier fort on the Great Plains in mid-western America in the 19th century. The frontier forts were places of safety in a wild landscape – some were just military camps, but others grew into busy towns, crowded with trading posts and workshops. Indians, fur-trappers, gold miners and farmers brought goods to the fort to sell them. Pioneers, on their way to set up new homes in the west, rested and bought provisions within the shelter of the fort's high wooden walls.

The forts needed the protection of high walls and other fortifications. From 1492, when Christopher Columbus landed, until late in the 19th century, America was a battleground. The new arrivals, from Spain, France, Britain, Holland, Portugal, even Russia, often fought each other. They also fought the Indians, or recruited different Indian tribes to help them in their battles. Then came wars between the United States of America, Britain and Mexico, followed by the American Civil War and the Indian Wars. In peace-time and in war, the frontier fort gave people a place to meet and rest in safety.

Tomahawk

Spear

Drum

Alaska

The Northwest Coast people carved huge dug-out canoes and beautiful totem poles.

The Rocky Mountains run north-south from Alaska to Mexico. Capped with glaciers and year-round snowfields, they divide North America in two. The Rockies' high peaks are home to mountain lions, grizzly bears and animals like bighorn sheep and mountain goats that are specially adapted to climbing steep slopes.

Harsh deserts lie between the Rockies and the sea. They include Death Valley, one of the hottest places on Earth. The deserts are famous for their cactuses, like the giant saguaro.

Totem pole

Elk

Wolf

Bald eagle

Rocky Mountains

Grizzly bear

THE NEW WORLD

The Europeans who settled North America called it the 'New World', but really the huge continent was just as old as Europe. In the late 1400s and 1500s, it was a land of snowy mountain ranges, dark forests, scorching deserts and endless plains. These teemed with plants and animals, many never seen before by Europeans. Grizzly bears and timber wolves hunted among the trees, and geese soared across the skies.

In 1492, when Christopher Columbus landed in the West Indies, he was welcomed by friendly Arawak people. Thinking he had arrived in India, Columbus called them 'Indians'.

Pacific Ocean

California

Mexico

Greenland

Pipe decorated with feathers

Hunting knife

Inuit

Seal

Arctic Ocean

The east coast forests were thick with deer and other animals. Tribes such as the Hurons and Mohicans hunted among the trees with bow and arrow, and gathered wild berries and nuts.

Two mighty rivers, the Missouri and the Mississippi, cut across the plains and flow into the Gulf of Mexico. Further north, the five Great Lakes join the St Lawrence River, which runs into the Atlantic Ocean. The cold northern rivers flow into Hudson Bay and the Arctic Ocean beyond.

Hudson Bay

Transport along rivers by birchbark canoe

St Lawrence River

Newfoundland

Beaver

The Great Lakes

The Great Plains (right) are endless stretches of tall grass where few trees grow.

Missouri River

The Great Plains

Virginia

Buffalo

Mississippi River

Atlantic Ocean

West Indies

Gulf of Mexico

The native people of North America belonged to more than 300 tribes and spoke many different languages. The richest and most powerful were the Aztecs. These fierce warriors ruled most of Mexico, sacrificing thousands of people from other tribes in bloody rituals every year.

Many smaller tribes were spread across the vast areas that are now called Canada and the United States. The most famous, like the Sioux (pronounced *See-oo*) and Blackfoot, lived on the open plains. Here they hunted deer, wolves and buffalo. They were nomads, packing up their teepees (skin tents) and moving camp with the changing seasons.

EXPLORERS AND SETTLERS

Dutch, Spanish, Portuguese, British, French – in the centuries after Columbus's discovery, people from every country in Europe came to North America. Some were looking for a new home; others came in search of gold and silver.

Columbus had been supported by Queen Isabella of Spain, and the Spanish were eager to follow in his footsteps. They had heard stories of *El Dorado*, a country where the streets were paved with gold. They travelled up and down the west coast of North America looking for this land of treasures. By 1776, 'New Spain' stretched as far south as San Francisco, California.

The Viking Leif Ericsson (above) was probably the first European to set foot on American soil. He landed in Vinland (Newfoundland, Canada).

The fortified village of Pomeioc (below), first seen by white settlers in 1585. The protective walls were made of tree trunks.

The Spanish *Conquistadores*, or 'conquerors' (right), crushed the Aztecs. By 1550, all of Mexico was part of 'New Spain'.

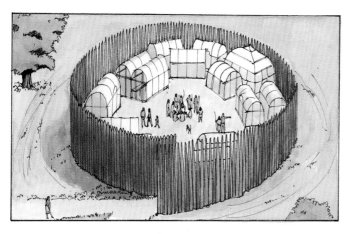

The first British settlers built Jamestown, Virginia (below) in the early 1600s. It was soon surrounded by huge tobacco plantations.

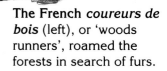

The French *coureurs de bois* (left), or 'woods runners', roamed the forests in search of furs.

New Amsterdam (below) was a Dutch fort and fur-trading centre. When the British captured it, they named it New York.

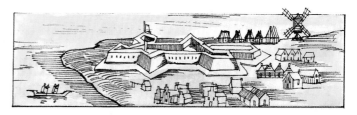

Fort Nelson, Canada
(above). This was one of
a string of fur-trading
posts built by the British
Hudson's Bay Company.

The Iroquois helped the
British fight the French.

**Like the majority of
American towns,** Charleston, South
Carolina (above) began
as a tiny walled village.

Forty years later
(below), Charleston had
grown into a bustling
port visited by ships from
all over the world.

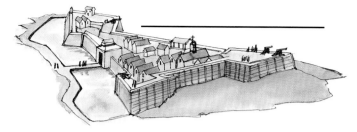

The French and English
fought a bloody war from
1754–63 (left). The
French had many strong
forts, but they lost the
crucial battle, on the
Plains of Abraham near
Quebec.

Pontiac (right), chief of
the Ottawas, led a
rebellion against the
British in 1763.

General Hamilton
(below), a British
commander, surrenders
Fort Sackerville to the
American Colonel Clark
in 1779 during the
American War of
Independence.

Trappers explored the woods and rivers, and
bought furs from the local people. The
French set up a chain of trading forts from
the St Lawrence River south to the Gulf of
Mexico.

English settlers landed in Virginia and
Massachusetts, on the east coast. The winters
were harsh, and the settlers would have
starved if the local Indians hadn't shown
them how to grow maize.

Two wars opened up North America. First
the British beat the French, taking control of
most of the land in 1763. Then, in 1775,
Britain's colonies rebelled. The War of
Independence ended in 1783 with the
foundation of the United States of America.

THE FIRST FORTS

For tired travellers trudging through the eastern forests, Fort Harrod in Kentucky was a welcome sight. In the 1780s, its high walls protected a small community of farmers. With simple hand tools, they had cleared the land and built the fort with the timber. Almost everything was made of wood; even the 'nails' were wooden pegs. A small spring inside the fort supplied all the water.

Farmers living in homesteads nearby came to the fort to grind their corn and buy supplies. Their children learned their lessons in a one-room school.

Lewis and Clark (left) were American explorers. They left St Louis, in Missouri, in March 1804 and headed west. With the help of Sacajawea, an Shoshini Indian woman who acted as guide and translator, they travelled across the Rocky Mountains. They finally saw the Pacific Ocean on 7 November 1805.

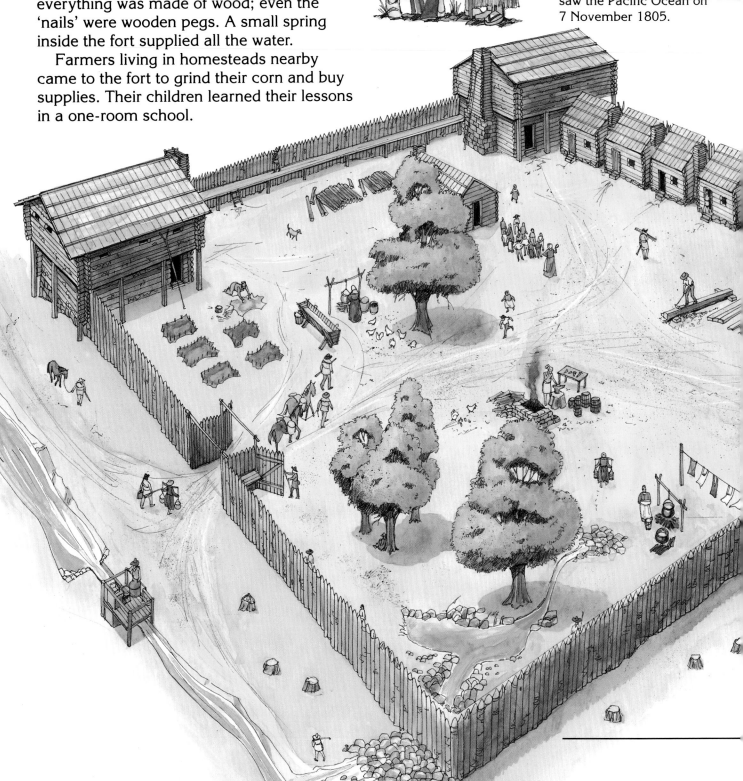

A 'mountain man' (above) dressed in home-made buckskin clothes rides into a fort with the body of an elk slung over his horse.

A fur trapper watches a log fort being built on the Yellowstone River (above). He is riding a half-wild Indian pony. Fur trappers were the first white men to explore the Rockies.

Like a castle in the Middle Ages, Fort Nez Percés (above) sheltered trappers in what is now Washington State.

The neighbours lend a hand to build a log cabin (left).

There were no roads through the forest, and the best way to travel was along the rivers by canoe (below).

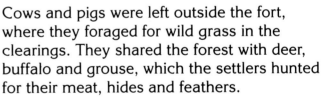

Below
Driven from their land by the new settlers, Cherokee Indians plod west on the 'Trail of Tears', 1838.

Cows and pigs were left outside the fort, where they foraged for wild grass in the clearings. They shared the forest with deer, buffalo and grouse, which the settlers hunted for their meat, hides and feathers.

Log cabins formed one wall of the fort. The leaders of the community lived in the bigger blockhouses at the corners. Guards with guns patrolled a raised walkway, keeping look-out for hostile Indians. The settlers cleared the land for 100 metres beyond the walls. This was further than a rifle could fire, so the settlers would have a little time to get ready.

If the fort was raided, everyone would run to shelter in the blockhouses. The stockade gates would quickly be closed and bolted with big sliding logs. The settlers fired their rifles through narrow windows, aiming at anyone they saw climbing over the wall.

THE WAY WEST

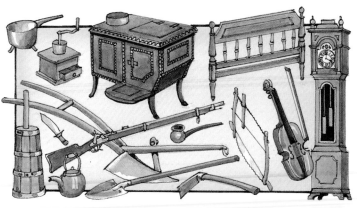

Their canvas tops flapping in the wind, a long line of wagons rolls across a sea of grass. This is a wagon train making the long trek west. From a distance, the wagons looked like sailing ships, which was how they got the nickname, 'prairie schooners'.

In the early 1800s, most of America was wild and unexplored. With no frontiers left to stop them, brave men, women and children packed up everything they owned and left their homes in the eastern United States to travel west and make a new life there. They became known as pioneers.

The pioneers had heard stories of the wonderful farming country in Oregon and California, on the Pacific coast. But to get there, they had to cross long stretches of dry prairie and desert, then drive their wagons up steep trails over the Rocky Mountains. The gruelling journey was over 3000 kilometres long and took from four to six months.

The pioneers usually travelled together because long 'trains' of wagons were less likely to be attacked by Indians. The biggest wagon trains included several hundred families, following each other in a line across the dusty plain.

The pioneers had to cram everything they would need on the journey and afterwards inside the wagon. Ploughs and shovels, a stove, a bedstead and a grandfather clock – it all had to fit in somehow. Often there was no room left for the people, who had to walk beside the wagon.

The 'prairie schooners' were usually pulled by teams of oxen. The wagons had no brakes or springs, so the ride was bumpy. Oxen are very slow walkers, so twenty kilometres was a good day's trek.

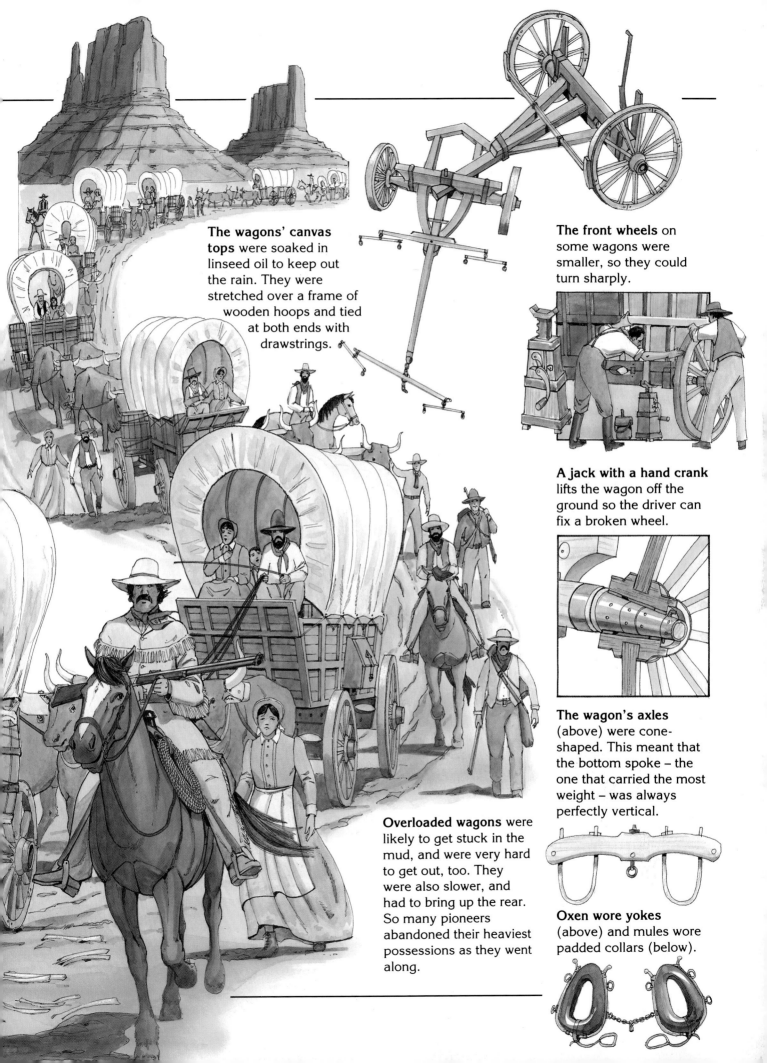

The wagons' canvas tops were soaked in linseed oil to keep out the rain. They were stretched over a frame of wooden hoops and tied at both ends with drawstrings.

The front wheels on some wagons were smaller, so they could turn sharply.

A jack with a hand crank lifts the wagon off the ground so the driver can fix a broken wheel.

The wagon's axles (above) were cone-shaped. This meant that the bottom spoke – the one that carried the most weight – was always perfectly vertical.

Oxen wore yokes (above) and mules wore padded collars (below).

Overloaded wagons were likely to get stuck in the mud, and were very hard to get out, too. They were also slower, and had to bring up the rear. So many pioneers abandoned their heaviest possessions as they went along.

THE OREGON TRAIL

Leaving their home in Boston, on the east coast, the family gets the train to St Louis. Then a steamboat carries them up river to Independence.

The family hears the cry 'Wagons, roll!' Their new wagon is fully loaded.

A sudden storm scares the oxen and blows the cover off the wagon.

It is May 1850, and a young family is about to set out for the green pastures of the Oregon country. They have left their home in the old city of Boston, Massachusetts, and are going to join a long wagon train assembling in the frontier town of Independence, on the Missouri River. Other pioneers they meet have travelled much further, from faraway countries such as Germany and Ireland. They are all going to follow the Oregon Trail.

There is much excitement in Independence as pioneers haggle over the price of a wagon and buy supplies for the long journey ahead. The wagon is so small – will it all fit inside?

The first part of the trail crosses great expanses of dusty prairie. The landscape gets drier and drier as the trail rises slowly towards the distant Rocky Mountains. Scouts ride ahead, keeping an eye on the cows and horses. A group of Indians comes down to trade some food with the settlers.

After two or three months, the route leaves the plain and cuts across the Rockies at South Pass. Some of the travellers are gold miners, off to try to make their fortune in California. Everyone calls them 'forty-niners', after 1849, the year that gold was first found.

The trail continues through the coastal mountains. It is autumn when the pioneers finally reach Oregon.

Breakfast is over. A bugle sounds, and the wagons set out again.

A broken wheel! Some fellow travellers help jack up the wagon and do a quick repair.

Fort Laramie, a rest stop. The tired pioneers buy fresh supplies. Father trades a knife for a deerskin from one of the Indians camped outside the fort.

There is no bridge over this river, so the wagons splash across.

Indians hunting buffalo stampede them past the wagon train.

As night falls, the pioneers camp in two big circles.

A passing buffalo hunter tells stories about the trail ahead.

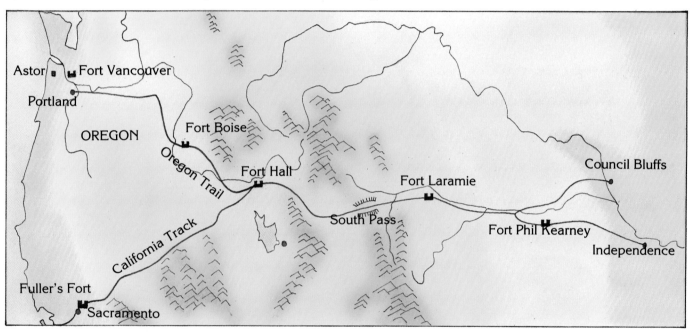

Astor • ■ Fort Vancouver

Portland

OREGON

Fort Boise

Oregon Trail

Fort Hall

Fort Laramie

Council Bluffs

South Pass

California Track

Fort Phil Kearney

Independence

Fuller's Fort

Sacramento

The family waves goodbye to friends going south to California.

The mountain paths are steep, and everyone has to push to help the oxen.

The wagons fan out in a dry valley. The trail is hard to find.

Oregon, at last! The wagons finally reach Fort Vancouver.

THE SITE FOR A FORT

The reconnaissance party surveys a possible site for the fort. They are looking for a flat spot on solid ground next to the river, close to the present Oregon Trail. The ideal place is a shallow bend where the river is easy to ford.

The engineer draws up a plan of the site (below).

The US government wants to build a new fort along the Oregon Trail. It will be a safe base for soldiers, who can protect the thousands of pioneers passing through every year. Army leaders consult with local trappers and an Indian agent. They decide to build the fort on a river near a buffalo range where Sioux and Comanche Indians camp and hunt in the spring and summer.

Work begins in the spring. It is a race against the changing seasons. The days are warm now, and living in a tent is fine. But will the new sleeping quarters be ready before the first snow?

A company of fifty soldiers and five officers arrives at the site. They make a detailed reconnaissance of the area and draw up plans. The river is a good source of fresh water. A pine forest to the north will provide all the timber. The soldiers also find limestone and red sandstone in cliffs to the south. This will be used for building the foundations.

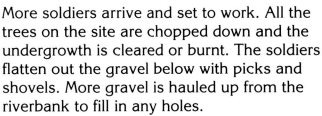

More soldiers arrive and set to work. All the trees on the site are chopped down and the undergrowth is cleared or burnt. The soldiers flatten out the gravel below with picks and shovels. More gravel is hauled up from the riverbank to fill in any holes.

At the same time, other soldiers cut and stack dry wood to use as firewood. Eight men do nothing but look after the horses.

The smaller trees are felled with axes. They will be used for timber or firewood.

The dense undergrowth is cleared away with scythes, the same tools that are used to cut hay.

Big trees are girdled. This means cutting away a ring of bark so the tree slowly dies.

Firewood is then piled around the dead tree and set alight. The huge trunk finally collapses.

A charred stump remains. The soldiers dig a pit around it, cutting the roots with axes.

A team of oxen is tied to the stump. Whipped into action, they drag it out of the ground.

The building site is full of activity (below). Once they have cleared the vegetation, soldiers break up the ground with picks and shovels. This is back-breaking work. Then teams of oxen drag harrows (heavy frames with metal teeth) up and down to level the soil. Until the stables are built, horses are kept in a makeshift corral.

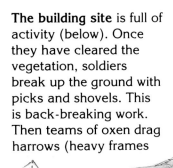

BUILDING THE FORT

Now that the site is level, work on the fort can begin. Companies of soldiers are busy cutting down trees. Teams of mules and oxen drag the heavy logs to the site. Here the trunks are trimmed and split to make good building timber. At the biggest forts, a sawmill was built to cut timber.

The fort is rectangular. The first buildings to go up are the blockhouses. There are three of these, two at opposite corners and a third above the main gate. These big buildings have a top storey, the balcony, which projects over the base.

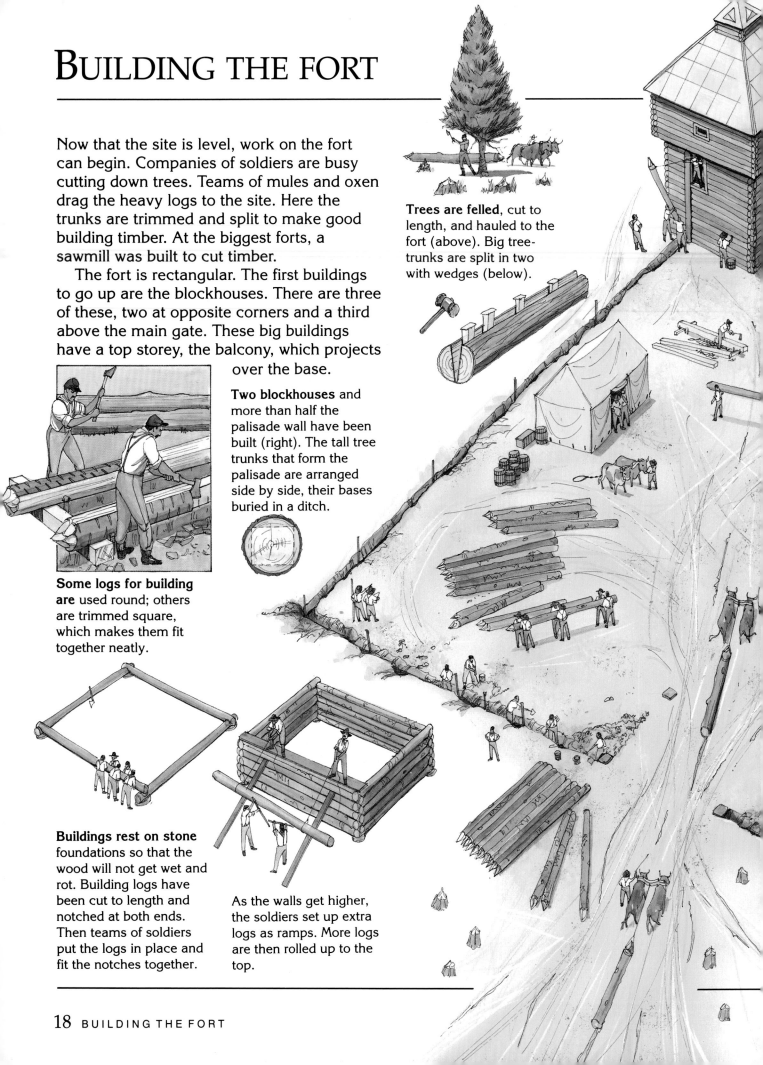

Trees are felled, cut to length, and hauled to the fort (above). Big tree-trunks are split in two with wedges (below).

Two blockhouses and more than half the palisade wall have been built (right). The tall tree trunks that form the palisade are arranged side by side, their bases buried in a ditch.

Some logs for building are used round; others are trimmed square, which makes them fit together neatly.

Buildings rest on stone foundations so that the wood will not get wet and rot. Building logs have been cut to length and notched at both ends. Then teams of soldiers put the logs in place and fit the notches together.

As the walls get higher, the soldiers set up extra logs as ramps. More logs are then rolled up to the top.

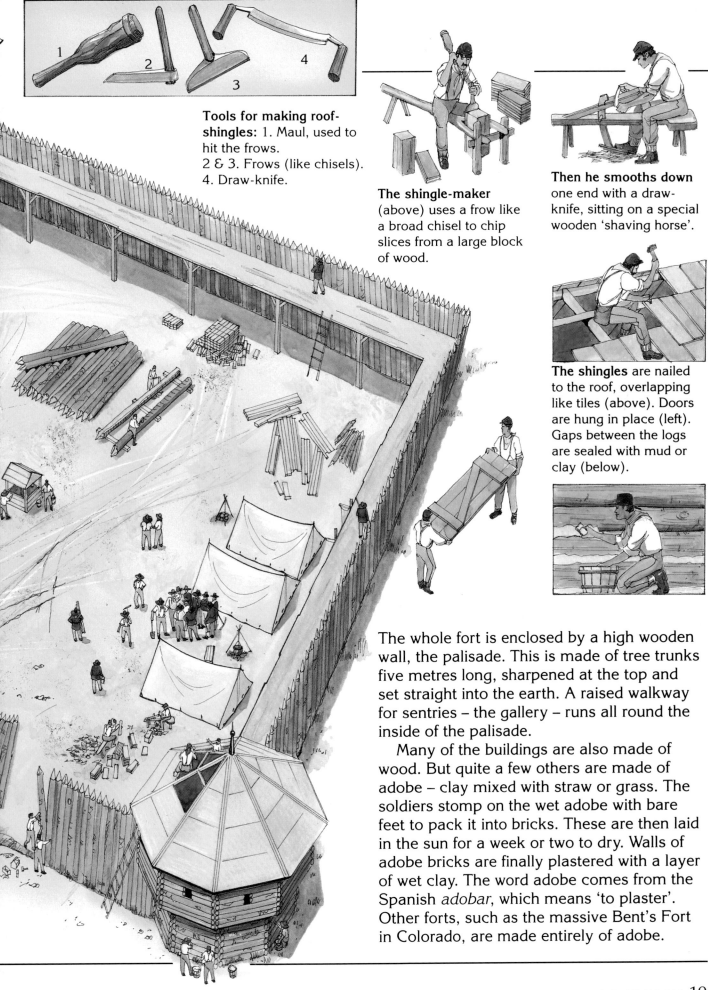

Tools for making roof-shingles: 1. Maul, used to hit the frows.
2 & 3. Frows (like chisels).
4. Draw-knife.

The shingle-maker (above) uses a frow like a broad chisel to chip slices from a large block of wood.

Then he smooths down one end with a draw-knife, sitting on a special wooden 'shaving horse'.

The shingles are nailed to the roof, overlapping like tiles (above). Doors are hung in place (left). Gaps between the logs are sealed with mud or clay (below).

The whole fort is enclosed by a high wooden wall, the palisade. This is made of tree trunks five metres long, sharpened at the top and set straight into the earth. A raised walkway for sentries – the gallery – runs all round the inside of the palisade.

Many of the buildings are also made of wood. But quite a few others are made of adobe – clay mixed with straw or grass. The soldiers stomp on the wet adobe with bare feet to pack it into bricks. These are then laid in the sun for a week or two to dry. Walls of adobe bricks are finally plastered with a layer of wet clay. The word adobe comes from the Spanish *adobar*, which means 'to plaster'. Other forts, such as the massive Bent's Fort in Colorado, are made entirely of adobe.

INSIDE THE FORT

The finished fort is a busy place. A company of infantrymen (foot soldiers) goes through their morning drill on the central parade-ground. Two privates stand by the big gun, a 12-pound howitzer, that faces the main gate. In the stables, cavalrymen feed and brush down their company's horses. The fort has a blacksmith and a saddler to shoe horses and mend saddles.

The soldiers' wooden barracks are heated with wood-burning stoves. The iron bedsteads stand in rows. The kitchen and mess-room are at one end of the building. There is no water on tap, so the soldiers fetch buckets from the well.

Most of the time, the Indians camped by the fort (above) got on well with the soldiers and the many passing settlers.

In spring and summer, wagon trains of pioneers come and go. The carpenter and wheelwright are kept busy repairing wagons and wheels. Pioneers, trappers and Indians all use the shop of the sutler, the fort's trader. They buy rice, tobacco, coffee, nails, and supplies of all kinds.

Sentries walk the walls, looking over the plain below. A band of Sioux Indians is camped just outside the walls. The Sioux come and go freely, collecting provisions and bringing furs and mocassins to exchange with the sutler – the fort's trader. They leave with pieces of colourful cloth, blankets, knives, ammunition and bottles of whisky.

Stables

Parade-ground

Sioux camp

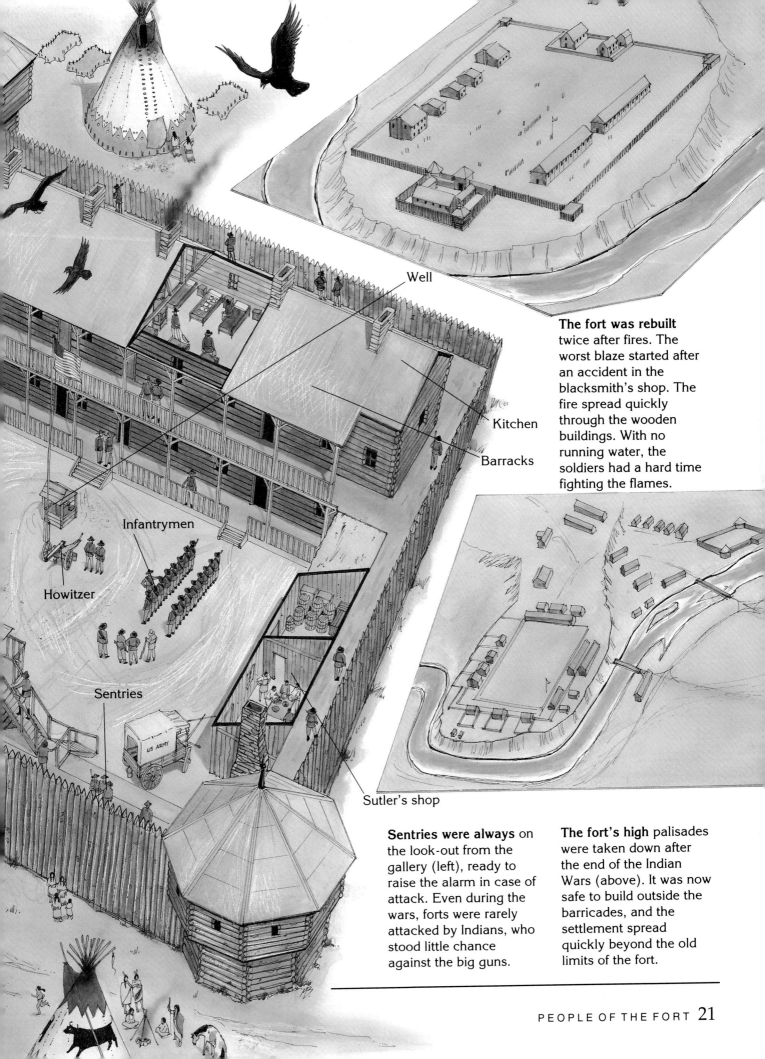

Well

Kitchen

Barracks

Infantrymen

Howitzer

Sentries

Sutler's shop

The fort was rebuilt twice after fires. The worst blaze started after an accident in the blacksmith's shop. The fire spread quickly through the wooden buildings. With no running water, the soldiers had a hard time fighting the flames.

Sentries were always on the look-out from the gallery (left), ready to raise the alarm in case of attack. Even during the wars, forts were rarely attacked by Indians, who stood little chance against the big guns.

The fort's high palisades were taken down after the end of the Indian Wars (above). It was now safe to build outside the barricades, and the settlement spread quickly beyond the old limits of the fort.

PEOPLE OF THE FORT

Captain

General

Sergeant

Private
(Infantryman)

Being a soldier in an American frontier fort was not very glamorous. The biggest forts, such as Fort Laramie, were like bustling frontier towns. But most were isolated outposts surrounded by unmapped wilderness. Visitors were rare, and life must have been very boring.

Senior officers enjoyed the most privileges. They often shared their houses with their families and servants. More junior officers could entertain friends in their private rooms, but privates (ordinary soldiers) lived together in barracks. The only women they were likely to meet were laundresses, or the general's wife! The only other civilians at many forts were Indian agents and visiting mountain men. Even the chaplain (priest) was employed by the army.

The captain was in charge of a company of 25 soldiers, but the sergeant kept discipline and gave the orders day-to-day. He supervised drills and inspections on the parade-ground. Senior officers such as generals and colonels would visit the fort from time to time, to pass on government orders, negotiate treaties with the Indians or organise attacks against them.

A private or regular infantryman ready for the day's duty (above). He is carrying a backpack with his tea cup dangling from one corner. This man has tied his rolled-up blanket on top of his backpack – other soldiers wear theirs across the chest.

Cavalryman on stable duty

Buffalo soldier

Cavalryman

Indian agent

Indian scout.

Infantrymen wore woollen uniforms and flannel shirts all year round. They sweated horribly in summer, and froze through the winter. Cavalrymen were allowed to wear canvas trousers, which protected the legs when riding. All soldiers had to pay for their own uniforms, so most bought cheap, faded clothes at auctions.

Ordinary soldiers were paid very little, but there was not much opportunity to spend money in a remote fort. Qualified men made a little extra money as barbers, blacksmiths or saddlers. Pay was handed out by the army paymaster, who visited the fort every two months. He carried a lot of money, and was always in danger from outlaws or Indians. So armed cavalrymen rode with him as he drove his wagon from fort to fort.

Cavalrymen were skilled horsemen who helped maintain order in the Wild West and fought in wars against the Indians. They wore knee-high leather boots, and often wore buckskin jackets instead of the regulation blue tunics. Many cavalrymen carried sabres (long, curved swords) as well as revolvers and rifles. When they were on stable duty, they dressed in white canvas clothes.

The Indian scouts often wore a strange mix of army uniform and Indian dress. They guided soldiers through the wilds and helped to trail other Indians or find their camps. Scouts were often from an enemy tribe; on the plains, for instance, Pawnee and Crow scouts helped in the fight against their traditional enemies, the Sioux.

FEEDING THE TROOPS

Army food was dreadful. The army hired very few professional cooks, so in most forts the soldiers took it in turns to do their own cooking. The food was often overcooked. The staples were stew, hash (cooked meat chopped up and re-heated), baked beans, salted meat (mostly pork), potatoes, melons and dried apples.

Meat was abundant around most forts. The plains and woods were full of game animals that could end up on a soldier's plate. The bigger forts also kept livestock and grew their own vegetables. The ranching and gardening were done by soldiers or hired help. Some forts had all sorts of problems keeping garden labourers, because they kept leaving their jobs for the gold rush in California. The problem was solved by hiring Mexican gardeners, who worked hard and had no desire to go searching for gold.

Supplies to feed the army and the constant stream of pioneers were brought all the way from the east. Special wagon trains left towns on the Missouri River and crossed the plains loaded down with foodstuffs. In 1858, for instance, 775 wagons pulled by nearly 8,000 oxen left the town of Atchison, Kansas, carrying over 3.7 million tonnes of supplies.

Meat was hung and salted (above) to preserve it, because ice was only available in winter from the frozen lakes.

Buffalo, elk, deer, jackrabbits and 'prairie chickens' (grouse) were all hunted. Cows, pigs and chickens were farmed.

Garden duty (below) included digging vegetables and feeding the chickens and turkeys. Gardens were usually near a river or creek, because watering the crops was a big problem on the dry plains.

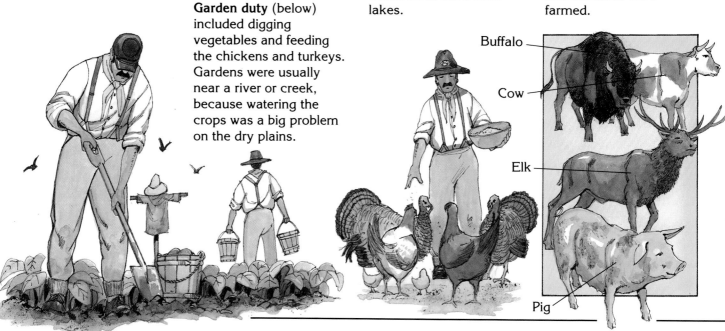

Buffalo

Cow

Elk

Pig

Corn bread   Hardtack

Hot coffee and pancakes smothered in molasses or maple syrup (top left) were served for breakfast. Melons (left) grew well on the plains, and were eaten as a snack or dessert. Cornbread, made from maize flour, was eaten with main meals such as stews, hash and 'white pot' (corn flour, milk, eggs and molasses). Another staple was hardtack – dry biscuits that tasted like cardboard. Fruit was eaten fresh, or dried.

Ordinary soldiers ate together in the canteen (above). Officers dined in the more refined atmosphere of the Officers' Mess, or if they were of senior rank, in their own houses, where servants waited at table. A lot of supplies were bought from local farmers (right), who were happy to sell their excess meat or crops to a nearby fort – it was cheaper and easier than shipping it back east. The kitchen (below) was always very busy.

A SOLDIER'S DAY

The soldier wakes up in the barracks at 6.30 a.m. He gets dressed, starting with long underwear called 'longjohns'.

A quick wash helps him wake up. In summer, the men often bathe in the river at the end of the day.

Dressed and ready. Every morning the soldier soaps his socks, so they won't give him blisters.

Lined up at attention on the parade ground. The sergeant assigns today's fatigue detail.

For an army private, life at the fort was a dull routine. Battles were rare, and the soldiers spent most of their time marching up and down the parade ground, doing hard manual work, or trying to stay awake on guard duty.

The soldier rose early. In winter, he was often washed and dressed before the sun was up. He was then assigned a fatigue detail – the day's work – by the sergeant. Fatigue details included working in the garden or kitchen, digging ditches, building roads or buildings, chopping firewood and fetching water from the river.

Soldiers always carried rations of coffee and hardtack in their backpacks. Out on

The soldiers have to repair a road washed out in heavy rain. They march 6 km to get there.

After the long march, the soldiers are digging in the hot sun by 9.00 a.m. The sergeant keeps watch.

A private's field sack (left) bulged with provisions for a day away from the fort. It held (1) tobacco; (2) matches; (3) a razor; (4) sewing kit; (5) coffee beans; (6) onions, potatoes and bacon; (7) hardtack biscuits; and (8) spare socks. The soldier also carried a tin water canteen (9), coffee mug (10) and a rolled blanket (11).

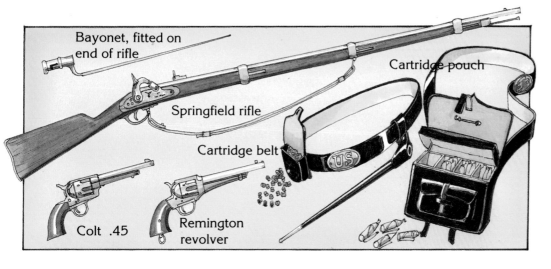

Bayonet, fitted on end of rifle

Cartridge pouch

Springfield rifle

Cartridge belt

US

Colt .45

Remington revolver

The main weapon (left) of the Western army was the Springfield Model 1873, a breech-loading, .45 calibre rifle. It had a long range and was very accurate. But it had to be re-loaded after each shot, unlike the repeating rifles carried by many Indians. Soldiers also carried a Colt or Remington revolver, both six-shooters.

Lunch break (below), a chance for the soldiers to relax and chat.

The march back (below). Until they reach the walls of the fort, everyone is watchful.

patrol, they often camped and cooked far from the fort. After a day in the saddle, cavalrymen complained that it was 'Forty miles a day on beans and hay.'

Soldiers enlisted for an initial term of five years. Bored or fed up with the routine, many deserted to try their luck at gold-panning or ranching instead. A visitor to Fort Laramie in 1850 wrote about eighteen soldiers who stole the best horses in the stables and rode off to try and make their fortunes in California. A command of cavalrymen was usually sent off after the deserters, but there was always the danger that the pursuers would decide to desert too!

The sergeant gets the men to chop some wood (below) before the evening meal.

Off duty (below), the privates play baseball on the parade ground.

After dark, cards and a drink in the traders' bar. It doesn't take long to lose a week's wages!

Sentry duty starts at midnight. His body aches, and only the cold air keeps the soldier awake.

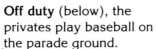

TRADERS, FURS AND GOLD

The large frontier forts were like islands in a vast wilderness, and they were often fur trading centres. Trappers who spent most of their lives alone in the wilds came to a fort to sell their year's furs. Traders working for big companies such as the Rocky Mountain Fur Company were waiting for them. So were sutlers, who sold blankets, food, whisky and supplies of every kind to all the fort's visitors.

The fur traders made huge profits from the trappers. But they often got an even better deal by exhanging furs with Indians. The Indians were fascinated by bright cloth and glass beads, and always wanted knives and ammunition. They had little idea of the relative value of these things. Unscrupulous traders tried to get the Indians to depend on fort supplies and alcohol, so they would keep coming back with furs and deerskins.

Pioneers, local farmers, gold miners and buffalo hunters also relied on the fort for news

White traders followed the army as it built forts in Indian country. This trader, in a suit and top hat, is smoking a pipe with some Sioux braves.

Gold miners, with heavily loaded mules (left), leaving the fort. Many miners crossed the country in search of the precious yellow metal. Most of them were badly equipped. Greedy traders made a fortune selling them tools and supplies at very high prices.

Skinning a buffalo (below). By 1870, professional hunters were killing 3 million buffalo a year. They only took the skins and tongues.

SMITH'S
HORSE SHOEING WAGONS
ALL FARM IMPLEMENTS REPAIRED

S. SMITH & SONS

and supplies. The first prospectors were the forty-niners on their way to California. Later, gold was found in other areas, such as the Black Hills of Dakota. Thousands of miners came to the West to try to make their fortunes.

The blacksmith's main job was shoeing horses, mules and oxen. Some smiths were soldiers who did most of their work for the army. Others set up shop outside the fort.

Smiths also made and mended tools such as ploughs for local farmers, and repaired knives and wagons for passing pioneers.

THE PLAINS INDIANS

After killing a buffalo, a brave thanks the Great Spirit. Then he eats the buffalo's heart to share in the animal's courage.

Indians lived in teepees made of buffalo hides (below). A fire was lit inside, and a smoke flap at the top was kept open for ventilation.

An Indian woman moves camp (above). She has used the poles from her teepee to make a *travois*, a carrying frame which is dragged behind a horse. Her young children sit on the *travois*, with the folded-up teepee cover. The woman has also bundled up her baby in a *papoose*, a deer-skin pouch.

Indian agents (right) worked for the federal government. They distributed any food or money the Indians had been promised in treaties Later they ran Indian reservations.

The Plains Indians had many meetings with government and army leaders. These often took place in a big teepee outside the walls of a fort. Before a conference could begin, everyone smoked the peace pipe.

The Plains Indians were nomadic – they moved their camps from place to place. In the summer, they set up camp near hunting grounds, moving to follow the wanderings of the buffalo. When the cold weather came, they packed up their teepees (skin tents) again and moved to a warmer winter camp, usually in a sheltered valley.

The buffalo hunt was the centre of Indian life. In the summer, the buffalo gathered in huge numbers. Sometimes the plains were black as far as the eye could see with thousands of dark woolly coats. When scouts located a big herd, the warriors (or braves) formed a hunting party and set off in hot pursuit. The women quickly packed up camp and followed too.

The hunters crept as close as possible to the grazing animals. Then with whoops and yells they rode through the stampeding herd, slaughtering as many buffalo as they could. The attack was very organised; some braves guarded the rear, while others maintained order in the ranks.

A great feast of raw or roasted buffalo meat was held to celebrate the hunt. Every part of the animal was used. Hides were tanned and made into clothes, blankets and teepee covers, and horns and bones were carved into tools and weapons. The extra meat was dried and stored for the long winter.

TRAPPERS AND TRADERS

The first white people to explore the western forests were fur trappers. In the first half of the nineteenth century, the great demand for beaver skins brought trappers to the unmapped areas around the Rocky Mountains. Many trappers were rugged people who were more at home in the wilderness than in a town or farm. They became known as mountain men.

Unlike the farmers and gold-miners who followed them, the mountain men usually lived at peace with the Indians. They adopted Indian ways, dressed in buckskin clothes and moccasin shoes, and learnt to speak Indian languages.

The life of a mountain man was lonely and dangerous. The social event of the year was the 'rendezvous'. This was a huge gathering of trappers, Indians and traders, usually in a flat valley with lots of room for everybody's teepees. The trappers came down from the mountains with their mules or packhorses loaded with furs. They exchanged these with traders for supplies and alcohol, and there was singing and dancing late into the night. Then the mountain men packed up and headed back to the wilds.

By 1840 beavers were rare. Many mountain men became guides for government expeditions or wagon trains.

A beaver trap is chained to a wooden stake on the bottom of a stream.

A beaver touches the bait pan and is trapped by the paw.

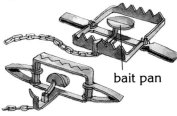

A beaver trap with jaws open and closed.

bait pan

Seeing an elk in the forest, the trapper shoots it for its meat and skin.

Trappers travel by birchbark canoe. The design is an Indian one.

The canoe is carried around waterfalls. This is called a portage.

A mountain man's Indian wife scrapes a buffalo hide (right). When it is clean of flesh, she will grease it, tan it and stretch it out to dry. She is already drying a deerskin and some strips of buffalo meat.

A trapper gives a bead necklace to an Indian, to encourage him to trade.

Indians and trappers in buckskin clothes tell tales around the campfire.

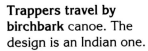

The rendezvous, a gathering of up to 2,000 trappers, traders and Indians.

Back at the campsite (above), the trapper cleans and presses his beaver skins. First he uses a metal scraper and a graining block to clean the flesh off the hide. Then he packs the pelts together to make a bale. He will hope to sell all the bales at the next rendezvous.

The trapper (below) wears a buckskin shirt, trousers and moccasins. His mule carries some furs and a dead elk. The trapper's Hawken rifle can kill a bear from 200 metres away.

INSIDE A PIONEER'S CABIN

Settlers on the plains could not find much wood, stone or clay, so many made their homes from blocks of turf, or sod. These sod houses were held together by the stringy grass roots in the soil.

The first sod houses were built by Mormon pioneers in the 1840s. Big blocks of sod were cut from the ground with a shovel or a special plough. Then they were laid like giant bricks and the cracks were filled with earth.

Plan of snake fence

Before barbed wire was invented, fences were made of wood or rocks. Here are a snake fence (above) and a rail and boulder fence (left).

Wheel

Tail

A windmill (above) pumped water from wells deep below the dry plains. The tail kept the wheel facing the wind. Farming was almost impossible without a steady water supply.

The simplest sod house (left, top) was cut out of a hillside. But there was always the risk that cattle would fall through the roof! An L-shape (left, centre) provided more shelter. When a farmer had saved up enough money, he could build a log cabin (left, bottom).

Walls two layers thick

Firewood was rare on the plains, so pioneers collected buffalo dung to burn in their stoves (right). They called it 'buffalo chips'.

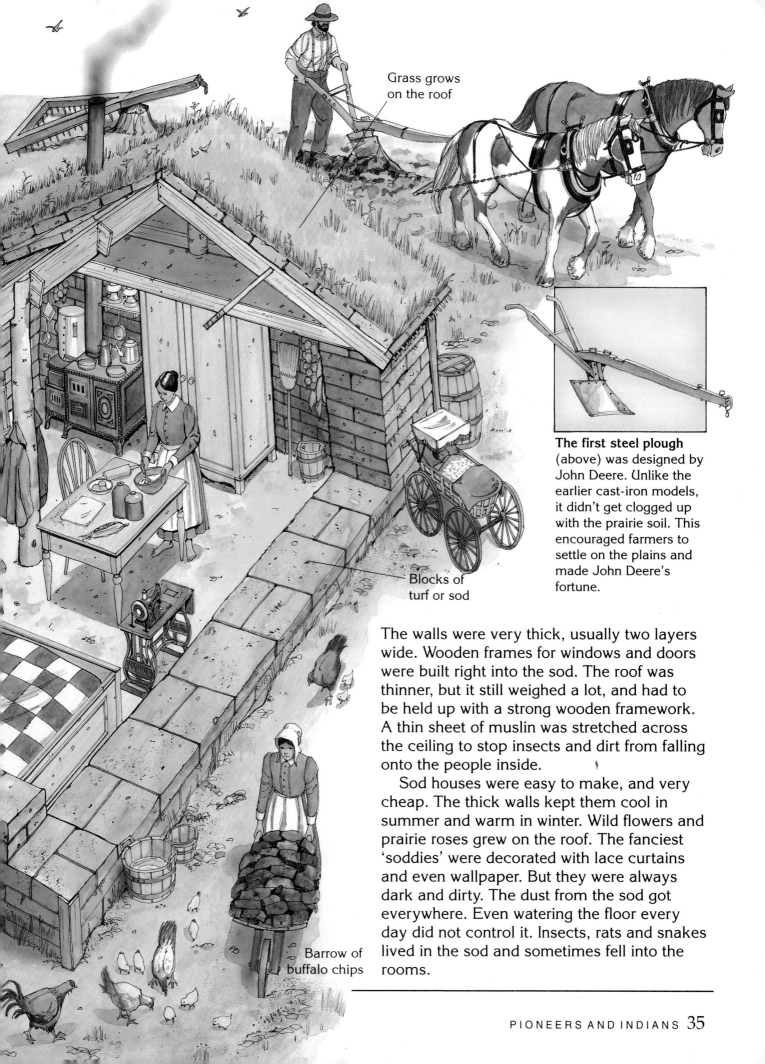

Grass grows
on the roof

Blocks of
turf or sod

Barrow of
buffalo chips

The first steel plough
(above) was designed by
John Deere. Unlike the
earlier cast-iron models,
it didn't get clogged up
with the prairie soil. This
encouraged farmers to
settle on the plains and
made John Deere's
fortune.

The walls were very thick, usually two layers
wide. Wooden frames for windows and doors
were built right into the sod. The roof was
thinner, but it still weighed a lot, and had to
be held up with a strong wooden framework.
A thin sheet of muslin was stretched across
the ceiling to stop insects and dirt from falling
onto the people inside.

Sod houses were easy to make, and very
cheap. The thick walls kept them cool in
summer and warm in winter. Wild flowers and
prairie roses grew on the roof. The fanciest
'soddies' were decorated with lace curtains
and even wallpaper. But they were always
dark and dirty. The dust from the sod got
everywhere. Even watering the floor every
day did not control it. Insects, rats and snakes
lived in the sod and sometimes fell into the
rooms.

A WOMAN'S DAY

Frontier women left the comforts of civilisation behind. Running a clean and organised home was a constant struggle, and they had to be hard-working and resourceful. Women spun their own wool and flax, and sewed clothes and bedding for the whole family. They cooked and cleaned, and learned how to make candles from buffalo or bear fat with wicks spun from plant fibres.

The woman of the house gets up at dawn and goes to the stream to collect water.

The next job is to milk the cows, which is done by hand and takes quite a long time.

The rest of the family is up and hungry for breakfast. Father has coffee, the children have fresh milk. The main course is pancakes with molasses (a type of sugar).

After breakfast the children help their mother feed the chickens and pigs with scraps.

The woman tends the vegetable patch while her children pick apples in the orchard.

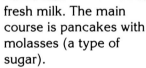

Some of the milk is stirred up in a churn, turning it into butter and buttermilk.

Before lunch, the woman finds time to bake some bread in her big iron stove.

She washes the family's clothes in the stream and hangs them up to dry between two trees.

The toilet is just an outhouse – a tiny wooden building over a deep hole in the ground.

The woman is sewing a doll for her daughter from scraps of left-over material.

Dinner, the main meal of the day, is roast turkey and vegetables followed by apple pie.

Neighbours help to husk the corn. They sit in the barn and talk while they work.

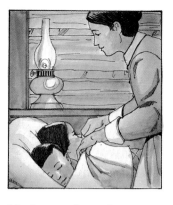

It's been a long day, and the children are exhausted. Their mother tucks them into bed.

A few minutes' peace to patch her husband's overalls, ripped while he was working in the fields.

A quilting 'bee'. Friends sew the quilt's three layers – the top one is a colourful patchwork.

Not all women kept house. Calamity Jane (right) dressed in buckskin and liked to drink and gamble with the men. Annie Oakley (below) was a sharpshooter in Buffalo Bill's Wild West Show (see page 45).

It was common for frontier women to have six or eight children. Many women died in childbirth, or lost children from diseases. It was hard to teach children the importance of cleanliness when water had to be fetched from a faraway stream and the main fuel was buffalo dung!

Another problem was loneliness, but when a family had a big job to do, like building a new barn or 'swingling' (threshing) their flax crop, they invited all the nieghbours round to help. When the work was done, the hostess would cook a huge meal, one of the farmers would play his fiddle and everyone would dance and sing.

Because they worked so hard, frontier women demanded more rights. Wyoming Territory led the way in giving women the vote in 1869, fifty years before all American women were granted that right.

BUILDING THE RAILWAY

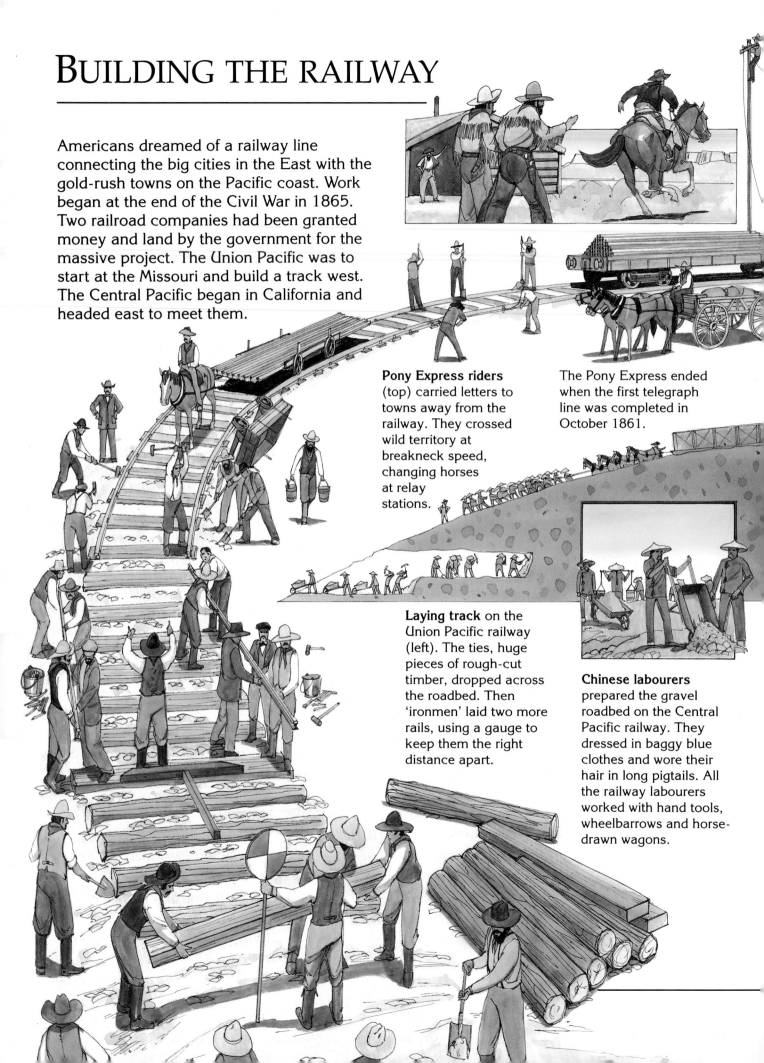

Americans dreamed of a railway line connecting the big cities in the East with the gold-rush towns on the Pacific coast. Work began at the end of the Civil War in 1865. Two railroad companies had been granted money and land by the government for the massive project. The Union Pacific was to start at the Missouri and build a track west. The Central Pacific began in California and headed east to meet them.

Pony Express riders (top) carried letters to towns away from the railway. They crossed wild territory at breakneck speed, changing horses at relay stations.

The Pony Express ended when the first telegraph line was completed in October 1861.

Laying track on the Union Pacific railway (left). The ties, huge pieces of rough-cut timber, dropped across the roadbed. Then 'ironmen' laid two more rails, using a gauge to keep them the right distance apart.

Chinese labourers prepared the gravel roadbed on the Central Pacific railway. They dressed in baggy blue clothes and wore their hair in long pigtails. All the railway labourers worked with hand tools, wheelbarrows and horse-drawn wagons.

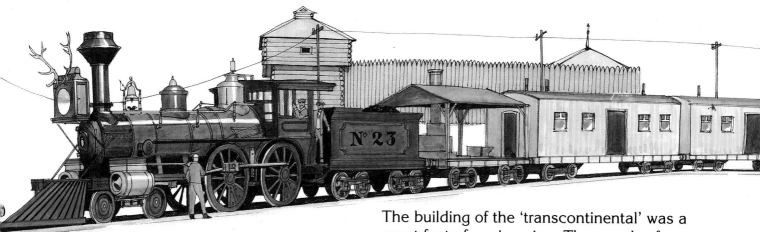

The Summit Tunnel (below) was 42 metres long. Workers blasted away the rock from both ends and from a shaft in the middle.

A small locomotive was used to raise the rock out of the shaft. Other supplies were dragged over the mountain so work could begin on the far side.

The building of the 'transcontinental' was a great feat of engineering. Thousands of labourers sweated in gangs with picks, shovels and gunpowder. The Union Pacific hired mostly Irish workers. Many had just arrived in America, or were veterans of the Civil War. Laying track on the flat plains was relatively easy. But the workers were often attacked by Indians, who were angry at the 'iron horse' (the train) invading their lands.

The Central Pacific railway had to cross the wild Sierra Nevada mountains. The company hired more than 12,000 Chinese labourers, most of whom were brought out specially from China. They proved to be hard and fearless workers. Tunnels and bridges had to be built through cliffs and across valleys, and the work was dangerous and slow. There were no bulldozers or drilling machines, and all the work was done by hand. Hundreds of workers died in accidents or avalanches set off by blasts.

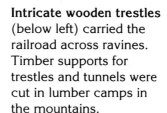

Intricate wooden trestles (below left) carried the railroad across ravines. Timber supports for trestles and tunnels were cut in lumber camps in the mountains.

Celebrations at last (below). On 10 May 1869, the two railway lines met at Promontory Point in Utah. The finished transcontinental railroad was 2,858 kilometres long.

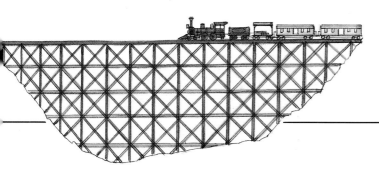

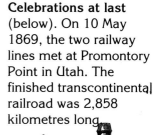

THE INDIAN WARS

The Sand Creek Massacre of 1864 (above). Soldiers killed at least 150 Cheyenne and Arapaho, mostly women and children.

Chief Joseph of the Nez Percé (above) made a famous speech when he surrendered.

General George Custer (above) was a controversial Indian fighter. He was killed in 1876, in the Battle of Little Bighorn.

Sioux Chief Red Cloud (above) fought the army when they tried to build forts on Sioux land in 1865. He won, and burnt the forts down.

Above
A group of Indians leave their land and set out for a reservation. When Chief Joseph finally surrendered, he spoke for all Indians: 'Hear me, my chiefs. I am tired. My heart is sick and sad. From where the sun now stands, I will fight no more forever.'

Geronimo (above) was an Apache chief. For 10 years he led his small band of warriors on raids in Mexico and Arizona, escaping from prison again and again.

Sitting Bull (above) was a Sioux medicine man and one of the leaders at Little Bighorn. He fled to Canada, but eventually surrendered. He was shot dead in 1890.

'The White Men were many, and we could not hold our own with them. We were like deer. They were like grizzly bears.' So said Chief Joseph at a council of the Nez Percé tribe in 1877. The US government had given his people thirty days to leave their home in Oregon and move to a reserve in Idaho. All over the frontier, the story was the same. Gold-miners, farmers and cattle ranchers were invading Indian land. They brought diseases that were unknown to the Indians. They slaughtered the buffalo and put up fences on hunting grounds.

The Indians fought many battles with the new settlers. In the 1860s the government tried to force tribe after tribe to leave their homelands and settle on reservations. Many Indians obeyed. Others chose to fight.

The Indians were brave and clever warriors. They rarely attacked the army in the open. Instead they made quick strikes or laid ambushes. But they were hopelessly outnumbered and were always running out of ammunition. The army also chose to fight in the winter, when the Indians were usually settled in their winter camps. Forced to flee from the soldiers, the Indians were often defeated by the snow and cold.

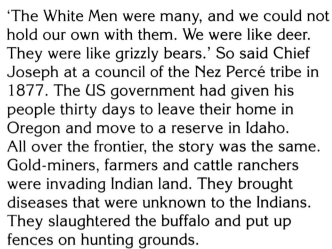

Indians believed the Ghost Dance (below) would make the white man disappear. The horrible massacre at Wounded Knee in 1890, when 300 Sioux were killed, proved them wrong.

Geronimo surrendered and moved to a reservation in 1886. The picture below shows him riding in a car in 1905.

A FRONTIER TOWN

A trip to town was a thrilling break from the boredom and loneliness of frontier life. As settlers poured into the West, bustling towns sprang up everywhere. Many started as forts and grew quickly into cities.

The shops had fake fronts, to make them look bigger than they really were. Many had porches with hitching-posts, where visitors could tie up their horses. The streets were bare earth, and the pavements were raised walkways made of wooden planks. On rainy days, pedestrians were splattered with mud by the wagons splashing down the street.

The West was full of 'boom towns' that seemed to grow overnight. Fort Worth, Texas was built by the army in 1849. Four years later, it had become a thriving trading centre, and the old fort was pulled down. The railway arrived in 1874, and the town was immediately overrun by cowboys and their herds of longhorn cattle. By 1885, Fort Worth had a population of 22,000.

Fort Worth was a 'cowtown'. Other towns 'boomed' when the goverment created new states and offered cheap (or even free) land to settlers. On the morning of 22 April 1889, there were four buildings in Guthrie, Oklahoma. By nightfall, 10,000 pioneers had set up their tents there. A council and mayor were elected the next day, and in a few weeks the town had a school, a bank, a church and its own newspaper, called the *Get-up*.

Meat is on sale at the town butcher's (below). Beef and pork come from local farms, and deer, rabbits, wild turkeys and ducks are bought from settlers who hunt them.

The General store (below) sells everything people need.

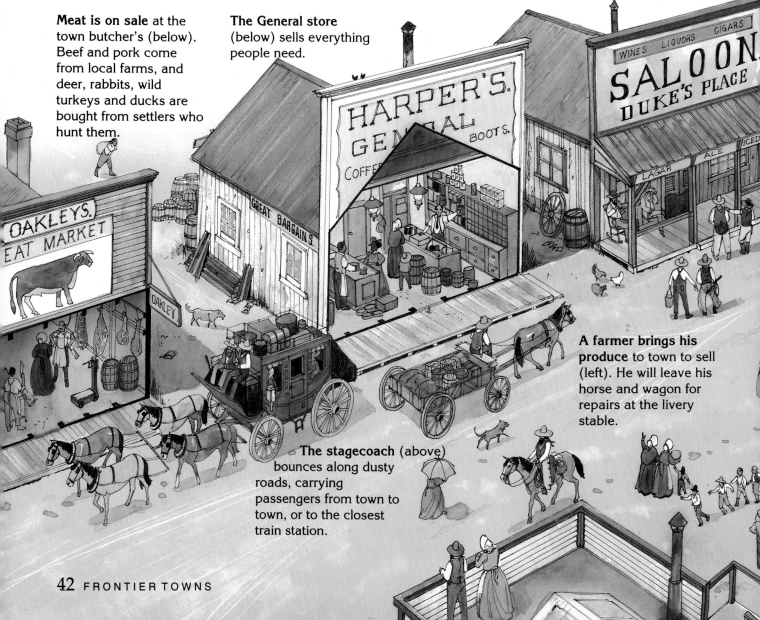

The stagecoach (above) bounces along dusty roads, carrying passengers from town to town, or to the closest train station.

A farmer brings his produce to town to sell (left). He will leave his horse and wagon for repairs at the livery stable.

Many settlers owe money to the town bank (below). Next door, the hotel rents cheap rooms. Its inn and dining rooms are a hub of town life.

The army post (behind) is a reminder of the town's military origin. The walls were pulled down years ago and the wood used to build houses.

Weddings, baptisms and funerals took place in the church (below). Many people only came to town on Sundays for the service there.

The town's water is pumped to the surface by a tall windmill (below). Anyone can fill their buckets or refresh their horses at its trough.

The doctor, or physician, runs the pharmacy (above). He also pulls teeth, amputates legs or removes bullets in his room upstairs.

FORT FACTS

The Alamo was a fortified mission in Mexico. It was captured in December 1835 by Texans, who were revolting against Mexican rule. Two months later, a huge Mexican force took the Alamo back. They killed every defender, including Davy Crockett and James Bowie. Soon afterwards the Texans defeated the Mexican army once again.

Bent's Fort had massive walls of adobe. It was the biggest fort west of the Mississippi and was the centre of the Colorado fur trade from 1833–48. Pioneers on the Sante Fe Trail stopped at Bent's Fort, but most of its visitors were fur trappers, traders or Indians.

Fort Laramie was an important stopover on the Oregon Trail in Wyoming. It was built in 1834, and soon became a major fur-trading centre. In 1849, the army bought Fort Laramie and turned it into a military post. An important treaty with the Sioux was signed there in 1867.

Fort Phil Kearney was built in 1866 in the Powder River Country of Wyoming. This was a traditional hunting ground of the Sioux, who were outraged. Their chief Red Cloud fought a clever war against the army. He won several victories and kept the fort under seige. The army finally agreed to leave the area. But Red Cloud would not sign the agreement until his warriors had burnt Fort Phil Kearney to the ground.

Fort Sumter was near Charleston, South Carolina. The first shots in the American Civil War were fired there on 12 April, 1861.

Fort Worth in Texas is just 50 km from the modern city of Dallas. Built in 1849, it soon grew into a thriving trading centre and city. It became a railhead in 1874.

Fort Bridger was built by the legendary trapper and guide Jim Bridger. The fort was on the Oregon Trail by Farewell Bend, where the California Trail leaves it and heads south. It was burnt to the ground by angry Mormons in 1857. The army rebuilt the fort, then abandoned it in 1890.

Quebec City is one of the only North American cities with stone walls that are still standing. The massive fortifications were built by the French in the 17th and 18th centuries. They weren't enough to defeat the British, who captured Quebec in 1759.

FAMOUS FRONTIER FOLK

Buffalo Bill Cody was a Pony Express rider, army scout, Indian fighter and buffalo hunter. He earned the nickname 'Buffalo Bill' after he killed more than 4,000 buffalo in one eighteen-month period. He is best known for his Wild West Shows, where real cowboys and Indians raced horses, performed riding and shooting tricks and enacted famous battles. The show even toured Europe, and performed for Queen Victoria in 1887.

Jim Bridger was a mountain man and army guide who was the first white man to see Great Salt Lake, Utah. He established Fort Bridger in Wyoming in 1843. He married three Indian wives, and outlived them all.

Billy the Kid was a cowboy, outlaw and murderer. His real name is unknown. In 1881, he was sentenced to death for the murder of Sheriff Brady, but he escaped from jail by killing two guards. He was eventually shot dead by Pat Garrett, an old friend who had become sheriff of Lincoln County.

Crazy Horse was a Sioux warrior. He set a trap for the army in 1866, when his braves killed more than eighty soldiers in the Fetterman Massacre. He was also one of the Indian leaders at the great victory of Little Bighorn. Unlike other Indian leaders, Crazy Horse never allowed photographers to take his picture. He was killed by a soldier in 1877.

Cochise was an Apache chief who fought a long war with the army in Arizona. He finally surrendered in 1871.

Kit Carson was a trapper, guide, soldier and Indian fighter. He fought in the Mexican War and the Civil War, and then lead ruthless campaigns against the Apaches, Navajos, Kiowas and Comanches. He died in 1868, from injuries caused by a fall from a horse.

Wild Bill Hickok was a gunfighter and sheriff. He claimed to have killed over a hundred men, but he was probably exaggerating. He survived several public shoot-outs, but was finally shot in the back and killed in 1877.

Annie Oakley was a crack shot and one of the stars of Buffalo Bill's Wild West Show. She could hit coins thrown in the air, and once shot a cigarette from the lips of the German Crown Prince. Sitting Bull called her 'Little Sure Shot'.

GLOSSARY

Adobe, built with bricks of clay and straw dried in the sun. Adobe buidings are then plastered.

Ammunition, bullets and gunpowder for firing guns.

Avalanche, a sudden slide of snow.

Barracks, building where soldiers sleep.

Bayonet, a long blade fitted to the end of a rifle.

Beaver, a large rodent that lives in the water. Beavers were prized for their fur, which was made into fashionable hats.

Bison, the correct name for buffalo.

Blockhouse, a two-storey building that forms part of the wall of a fort.

Brave, a young Indian warrior.

Buck, a male deer. The skin of a buck used to be worth one dollar, or 'buck'.

Buckskin, tanned deer hide used to make clothes like buckskin jackets and moccasins.

Buffalo, a large wild cow with thick, dark fur. Buffalo once roamed the plains in huge herds. The correct name is American bison.

Buffalo soldier, a black soldier. They were given their name by Indians, who thought their dark, curly hair looked just like a buffalo's.

Bugle, a small trumpet.

Colt, a famous gun manufacturer. The Colt .45 was one the most common revolvers on the frontier.

Conquistadores, the Spanish 'conquerors' or soldiers who invaded South and Central America in the 16th century. They destroyed the Aztec, Mayan and Incan civilizations.

Coureurs de bois, French 'woods runners' who collected furs in the northern forests. They got on well with the Indians.

Detail, a soldier's special task or duty.

Elk, a large deer with impressive antlers.

Fatigue, a non-military duty for a soldier.

Fell, to cut down a tree.

Ford, to cross a river by wading across.

'Forty-niners', gold-miners who rushed to California in 1849, hoping to make their fortune there.

Frontier, the wild lands to the West in 19th-century America. The frontier was always being moved west as settlers fenced and farmed the land and cities grew.

Gallery, a raised walkway inside the palisade of a fort, from which sentries can keep watch.

Gauge, the width of a railway track.

Grouse, a plump bird that can barely fly. Grouse are very good to eat, and were often called 'prairie chickens'.

Howitzer, a large, powerful gun moved around on wheels.

Indian agent, a government worker who distributed supplies to Indians and dealt with their complaints.

Mess-room, an army dining room.

Moccasins, shoes made of tanned deer- or moose-hide.

Molasses, thick, sugary syrup. Also called treacle.

Mule, a cross between a horse and a donkey.

Officers' mess, a special dining and living room for army officers.

Ox, a large, strong cow used for heavy work like pulling wagons.

Oxen, the plural of ox.

Palisade, a high wooden wall made of upright logs.

Papoose, a Indian buckskin pouch for carrying a baby. The papoose could be carried on a person's back or strapped to a horse.

Pioneer, a settler in the Wild West.

Portage, carrying a canoe around rapids or a waterfall or between two rivers.

Prairies, the flat grasslands that cover large expanses of the West.

'Prairie schooner', a covered wagon.

Private, a common foot soldier of the lowest rank.

Reconnaissance, a detailed survey of an area, usually conducted by the army.

Reservation, an area set aside for Indians to live in. Many reservations are on poor land.

Revolver, a small gun held and fired with one hand.

Sawmill, a factory where logs are cut up into timber.

Scouts, guides who helped travellers or soldiers find their way around or locate Indian camps. Many scouts were Indians or retired mountain men.

Sentry, a soldier ordered to keep guard on a fort or expedition.

Shingles, large wooden roof tiles.

Spoke, a supporting rod that runs from the hub to the rim of a wheel.

Stampede, a sudden scattering of horses, buffalo or other animals.

Sutler, a fort trader.

Teepee, a portable tent. The Plains Indians made teepees from buffalo hide; the Northeast tribes used tree bark.

Totem pole, a log, carved with faces and animals, put up in front of a house. The Indians of the Northwest coast still make totem poles.

Travois, a wooden carrying-frame dragged by a horse or dog. Indians made their travois from the main posts of their teepees.

Trestle, a bridge made from an intricate framework of wood.

Wagon train, a big group of wagons travelling together, usually in single file.

Well, a hole in the ground from which water is lifted or pumped.

Yoke, a wooden collar fitted over the neck of a working animal like an ox.

INDEX

Page numbers in bold refer to illustrations.

A

adobe 19, 44, 46
Alamo 44
Arawak 6
Aztecs 7

B

barracks 20, 46
Bent's Fort 44
Billy the Kid 45
blacksmiths 29, **29**
blockhouse 11, 18, **18**, 46
buffalo 29, 30, 46
Buffalo Bill Cody 45
'buffalo chips' 34, **35**

C

Calamity Jane 37, **37**
California 12
canoes 32, **32**
Carson, Kit 45
cavalrymen **23**
Central Pacific railway 38
Charleston, South Carolina 9,
    9
children 37
Chinese labourers 38, **38**, 39
Columbus, Christopher 6
Conchise 45
conquistadores 8, **8**, 46
coureurs de bois 8, 46
cowboys 42
Crazy Horse 45
Custer, General George 40,
    40

D

Death Valley 6
Deere, John 35

E

El Dorado 8
English settlers 9
Ericson, Leif 8

F

fences **34**
food 24-25, **24-25**
fort, building a 16-17, **16-17**,
    18-19, **18-19**
Fort Bridger 44, 45
Fort Harrod 10
Fort Laramie 14, 22, 27, 44
Fort Phil Kearney 44
Fort Sumter 44
Fort Worth 42, 44
French settlers 9
fur trading 28
fur trappers 11, **11**, 32

G

gardening 24
General Hamilton 9, **9**
Geronimo 41, **41**
gold miners 14, 29, **29**, 46

H

Hickok, Wild Bill 45
houses 34-35, **34-35**
hunting 29, 31

I, J

Indians 6, 11, 14, 15, 20, 28,
    28, 30-31, **30-31**, 32, **33**,
    39
Indian wars 40-41, **40-41**
infantryman 22, **22**, 23

Jamestown, Virginia 8, **8**

L, M

Lewis and Clark 10, **10**
Little Bighorn, Battle of 40,
    45

Mexico 7
Mormons 34
mountain men **11**, 32
mules 13

N

New World 6
New York 8, **8**
North America 6-7
    map of 6-7
    native people of 7

O

Oakley, Annie 37, **37**, 45
Oregon Trail 14-15, 16
    map of 15
oxen 13

P

palisade **18-19**, 19, 21, 47
pioneers 12, 14, **14**, 16, 20,
    47
Plains Indians 31
Platte River 15
plough, steel 35, **35**
Pomeioc 8, **8**
Pontiac **9**
Pony Express 38, **38**, 45
provisions 26, **26**

Q, R

Quebec City 44
Queen Isabella of Spain 8

railway 38-39, **38-39**, 42
Red Cloud, Chief 40, 44

rendezvous 32, **32**, 33
reservation 40, 47
Rocky Mountain Fur
    Company 28
Rocky Mountains 6, 12, 14,
    32

S

Sand Creek Massacre 40
scouts 14, 23, **23**, 47
sentries 21, 47
settlers 34
shingles, making 19, **19**
Sioux Indians 20, **20**
Sitting Bull 41, **41**
soldiers 16, 17, 18, 19, 20,
    22-23, **22-23**, 26-27, **26-
    27**, 40
stagecoach 42, **42**
steamboat **14**

T

teepees 30-31, **30-31**, 33,
    47
telegraph 38
tools **19**
town, frontier 42-43, **42-43**
'Trail of Tears' 11
traders 28, **28**
trappers 28, 32, 33, **33**
travois 30, **30**, 47

U

uniforms 23
Union Pacific railway 38

W

wagons 12-13, **12-13**, 14-
    15, **14-15**, 20
war 9
weapons 27, **27**
windmill 34, **34**, 43
women 36-37, **36-37**

DATE DUE

9/17/09			